Contents

Executive Summary 2

Introduction 5

I. Technical and Economic Considerations in Renewable Integration 7

    Characteristics of a Grid with High Levels of Variable Energy Resources 7

    Technical Feasibility and Cost of Integration 12

II. Evidence on the Cost of Integrating Variable Renewable Generation 15

    Current and Historical Ancillary Service Costs 15

    Model Estimates of the Cost of Renewable Integration 17

    Evidence from Ancillary Service Markets 18

        Effect of variable generation on expected day-ahead regulation mileage 19

        Effect of variable generation on actual regulation mileage 22

        Implications for Smart Markets and Storage 23

III. Expanding Opportunities for Emerging Technologies 26

    Smart Markets and the Smart End-user 26

    Energy Storage 30

IV. Conclusions 34

References 35

Appendix 39

Executive Summary

The cost of renewable energy has been quickly dropping and renewable energy generation has been rapidly growing in the United States, spurred by state and federal policies and technological advances. Moreover, projections going forward suggest ever increasing penetration of renewables into the electricity grid. The two most rapidly growing renewable energy sources, wind and solar, provide variable energy output that depends on the time of day, location, season, weather, and other factors. Integrating high levels of these renewables onto the grid will require a reimagining of the management of the grid. It will increase the demand for grid management services, opening up a new set of important opportunities for promising technologies and approaches. This report examines economic and technical considerations related to increasing integration of variable renewable energy resources onto the existing electric grid, which highlight the importance of emerging technologies and approaches in smart markets and energy storage that can help smooth this transition. Smart markets use new communications technologies to develop integrated approaches allowing for electricity demand to respond during times of high value. Energy storage technologies allow the temporary storage of electricity so it can be released during times of high value. The key report findings are outlined below.

Renewable electricity from wind and solar is rapidly growing around the world.

- Wind and solar are known as "variable energy resources" (VERs) because their output is variable. Generation of wind and solar depends on when the sun is shining and the wind is blowing, which is imperfectly predictable.
- There are regions in the world that are already successfully managing an extremely high penetration of renewable VERs. For example, Portugal was run 100 percent on wind, solar, and hydropower for four days straight in May 2016, and Texas hit a record level of 45 percent instantaneous penetration from wind generation during one evening in February of this year.

The distinctive characteristics of VERs will likely require a reimagining of electricity grid management.

- One characteristic of a system with a high penetration of solar energy is the possibility that net electricity load (i.e., electricity demand minus VER generation) may exhibit a "duck curve." A duck curve, as seen below in the 2020 forecast for the net load during a typical spring day in California, contains a steep ramp of net load downward in the morning as the sun rises and a steep ramp upwards in the evening as the sun sets.

Net Load for March 31

Source: CAISO

- The steeper net load ramps seen in the duck curve, which can also occur more frequently in a wind-dominated system, as well as greater uncertainty in net load forecasts from the VER generation, are expected to raise the value of grid management services. These services ensure that there is sufficient capacity and that the voltage and frequency on the grid are within a narrow range.
- In the past, many energy analysts and academics had argued that the variable nature of these sources would limit their integration onto the grid. As this report demonstrates, though, there are areas with extensive use of renewables and many new technologies can help meet the demand for grid management, allowing for high levels of renewable VERs.
- Estimates of current grid management prices at the current levels of renewable VER generation across different electricity regions show no obvious relationship between the level of VER penetration and costs so far. Other factors, such as natural gas prices, have thus far been more important than VER generation for driving grid management service prices, but this may change with higher levels of penetration.

Estimation results show that increased steepness in VER generation is associated with greater demand for grid management services.

- Using data from California, a large electricity market with a higher penetration of renewable VERs than most, we find that the grid operator uses more grid management services during times of greater ramping in both electricity demand and renewable VER generation. In particular, solar VER generation has an important impact on the use of certain grid regulation services.
- We find that even at today's penetration of renewables, a service that could entirely flatten the steepest one hour of ramp of VER generation can save just under $900,000 in grid management costs annually in California, or up to $6.6 million annually based on higher ancillary services costs in the mid-Atlantic region. This is only one of several possible services

that could be provided by developers of new technologies and approaches that can help facilitate integration.

The increased demand for grid management services offers growing opportunities for smart markets and energy storage.

- New technologies and approaches hold promise to allow for a smart end-user working within a smart market. "Smart markets" refer to communications technologies and approaches that facilitate end-user responses in the demand for electricity. Such technologies and approaches can allow electricity demand to more easily adjust during times of stress on the grid.
- Notable declines in the price of energy storage technologies are fostering an emerging market. These energy storage technologies can temporarily store electricity for later use during high value periods. For example, lithium-ion battery packs used in storage systems have dropped in price from over $1,000 per kWh to under $400 per kWh since 2007.
- The promise of these new technologies and approaches can only be met if markets are structured in a way to allow these new technologies and approaches to provide grid management services. Smart markets and energy storage need to be able to bid into different grid management markets in a way that captures their full value.
- Recent action to update regulation at the state and federal level will help these promising new technologies and approaches capture further value streams and provide important services to the grid. Recent orders by the Federal Energy Regulatory Commission (FERC), such as Orders 755, 784, and 792, are important for allowing these new technologies and approaches to participate in electricity markets and capture the value they provide to the grid. However, not all regions are subject to FERC orders, and there is further room to enable participation of distributed resources in power markets.

Introduction

The United States is undergoing a dramatic transition to an electric grid powered substantially by renewable generation that will help reduce emissions and lessen future impacts from climate change. Federal and state policies have contributed and continue to contribute to this transition. The Recovery Act was the largest investment in history in clean energy, spurring private sector innovation and investment in new technologies around the country.[1] The Clean Power Plan, production tax credit, and investment tax credit all promise to continue facilitating further growth in renewable energy markets in the coming years. For many years, state renewable energy policies have successfully deployed renewable energy across the country. In addition, President Obama has joined the leaders of 19 other countries and the European Union, representing over 75 percent of the world's CO_2 emissions from electricity, in "Mission Innovation," a commitment to double clean energy research and development over the next five years.

Wind and solar—the most rapidly expanding renewable electric generation resources—are known as "variable energy resources" (VERs) because their output occurs when the wind blows and the sun shines, and cannot be controlled by the electric grid operator.[2] These zero-carbon resources will help reduce emissions and support energy independence, and are growing rapidly. Wind and solar generation accounted for 5.3 percent of total U.S. electricity generation in 2015, more than doubling from under 2 percent in 2009.[3] Moreover, they are expected to continue growing rapidly, especially with considerable price declines of 80 percent for utility scale solar and 60 percent for wind since 2009.[4] As the penetration of renewable VERs reaches much higher levels, their distinctive features will require a re-envisioning of the management of the grid. For example, some energy analysts and academics have long argued that wind and solar combined can never reach more than 15 to 20 percent penetration into the grid due to the need to meet demand at all times and ramp up and down other resources to balance intermittent ones.[5]

However, we already see evidence of much greater penetration of renewable VERs in some regions. For instance, Texas hit a record level of instantaneous penetration from wind generation during one evening in February of this year, with wind energy providing 45 percent of total

---

[1] CEA (2016).
[2] Technically, VERs can be "turned-off" by the grid operator or even withheld to be made available when needed. Both of these options are costly.
[3] EIA (2016a).
[4] Malik (2015), Stark et al. (2015). For some examples of the dramatic increases in renewable energy penetration that can be expected, Hawaii enacted a law in 2015 that set a goal of 100 percent renewable energy by 2045, following on its goal of forty percent by 2030 (Hawaii State Energy Office 2016). California's Renewable Portfolio Standard (RPS) requires 33 percent renewable energy by 2020 and a bill passed last year sets a goal of 50 percent renewables by 2030 (CPUC 2016, CEC 2016). Colorado's RPS requires 30 percent of electricity from investor-owned utilities from renewable resources by 2030 (Colorado 2016). New York also has a goal to generate 50 percent of its electricity from renewable resources by 2030 (NYSERDA 2016).
[5] Farmer (1980), Cavallo (1993), DeCarolis and Keith (2005).

electric power.[6] Internationally, penetration of variable electricity as a percent of total annual electricity consumption reached high levels in many countries in 2015, such as 42 percent from wind in Denmark, 24 percent from wind in Ireland, and 23 percent from wind in Spain with instantaneous generation rates at times exceeding 100 percent of electricity usage in some of those nations.[7] Just earlier this year, Portugal ran for four days straight on 100 percent renewables (wind, solar, and hydropower).[8]

Such higher penetrations of renewable VERs today are often made possible by large connected grids and high levels of hydropower capacity. But they also come with increasing demand for grid management services (i.e., services provided by generators or others to ensure reliable electricity provision) to facilitate the integration of renewable VERs. Such increased demand is creating a new set of important opportunities for promising technologies and approaches. Foremost among these promising approaches are energy storage technologies and smart electricity markets that enable demand to be flexible and respond when needed. These technologies and approaches are rapidly improving and can be enabled by a regulatory structure that levels the playing field. Further, these technologies and approaches can capture growing value streams—avenues that can be monetized by developers—and open opportunities for improved grid management that are only all the more important with increased renewable penetration.

This report reviews the economics of electric grid operation and integration of renewables into the grid, highlighting the promising opportunities for emerging technologies and approaches that allow for smart markets and energy storage. The next section provides a discussion of the fundamental aspects of the operational and economic considerations inherent in the adoption of high levels of renewable electricity generation. Section 3 analyzes grid operation to demonstrate how renewables are currently being integrated into the grid and uncover some of the value streams that can support the broader deployment of smart markets and energy storage both now and in the future. Section 4 discusses the opportunities from emerging technologies and novel applications of storage and demand response to support the management of the future electric grid.

---

[6] ERCOT (2016), UtilityDive (2016).
[7] Irish Wind Energy Association (2016), Energinet.dk (2016), RED Electrica de Espana (2016).
[8] APREN (2016).

I. Technical and Economic Considerations in Renewable Integration

Increasing the level of renewables on the grid raises both technical and economic questions. First, how does the grid operator manage generation from renewable VERs? Second, what does this imply for the technical feasibility and cost of integrating higher levels of renewables onto the grid? This section answers both of these questions by first discussing what integrating renewables onto the electricity grid entails from a technical perspective and then discussing the cost of integration to the electricity grid, setting the stage for the later discussion of the emerging opportunities for smart markets and energy storage.

Characteristics of a Grid with High Levels of Variable Energy Resources

The electricity market has some similarities to most other commodity markets in that wholesale prices are set by equilibrating the supply and demand for electricity. One difference from most markets, however, is that the supply and demand of electricity must be matched at every moment within a limited band of frequency and voltage. Inventory is only possible with energy storage, which is site-specific, resource limited, and historically has played only a minor role in electricity markets. Second, the demand for electricity is not perfectly known in advance and retail prices for consumers are not time-varying under most utility rate plans in the United States. If they are time-varying, they tend to have a simple schedule with one price at night and one price during the day that does not vary as wholesale market conditions change. Thus, with no price incentive to shift the timing of usage, the demand for electricity is highly inelastic even over long time frames and perfectly inelastic over short time frames. This means that advance planning is essential in a way that differs from most markets. Although the operation of the grid varies by electricity region, large areas of the country are managed by regional transmission organizations that coordinate the dispatch of electricity from generators, often through markets for commitments to generate at a given time in the future. Depending on the regulatory structure of the market, the generators themselves may be utilities and/or private companies that own generation capacity.

There are several unique characteristics of electricity output from the most rapidly expanding renewable resources (wind and solar) that differ from most other electricity generation. First, wind and solar generation is not *dispatchable*, which means that the grid or facility operator cannot choose when to dispatch the electricity from the facilities.[9] Only when the sun shines or wind blows can electricity be generated. Second, wind and solar generation generally exhibits cyclical variations in output based on the daily fluctuations in wind speeds and solar irradiance. Although it varies by location, wind generation is often greatest in the very early morning hours of the day, and solar generation is often greatest around midday and early afternoon. Third, both wind and sun exhibit minute-to-minute changes in wind speeds and solar irradiance. This implies

---

[9] Technically, VERs can be dispatchable in the sense that they can be turned off or "curtailed" when they are producing. But they cannot produce more electricity when needed unless they are already curtailed and curtailment is wasteful because it means available nearly zero-cost electricity is unused.

that the actual amount of production at any given moment in time from a wind or solar facility may be difficult to predict in advance. Fourth, the short-run cost of providing an additional kilowatt-hour (kWh) of wind and solar generation—provided that the wind is blowing and sun is shining—is nearly zero.[10] Wind gusts and solar irradiance are free and the fixed costs of installation must be paid regardless of the amount generated by the facility.

Figure 1 provides an illustrative example of the cyclical and non-cyclical variations of wind and solar photovoltaic generation on the Electricity Reliability Council of Texas (ERCOT) grid, which covers most of Texas except the panhandle. The figure shows system-wide centrally-generated electricity generation at 5 minute intervals for an illustrative day (the figure underestimates variability that occurs at individual-level sites or at time intervals shorter than 5 minutes). Notably, solar generation ramps up during the middle of the day, while much of the wind generation is at night. Solar generation also tends to have a steeper ramp up in the morning and down in the evening than wind. Both solar and wind generation display many short-run fluctuations throughout the day.

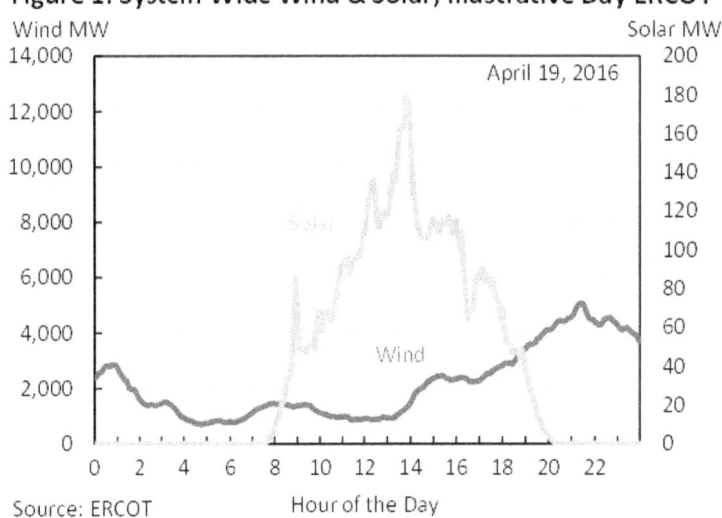

Figure 2 provides a similar figure for the generation of a variety of renewable energy technologies over an illustrative day in March 2016 in the California Independent System Operator (CAISO) electric grid, which covers nearly all of California. Since this graph shows average hourly output, the very short-run fluctuations that occur at the 5 minute level are smoothed out, but the broader cyclical trends are quite apparent. Wind generates the most electricity at night during this particular day, while solar generates the most in the day. While this negative correlation may differ by location and season, when it occurs, it is one advantage of pairing the two technologies. In contrast to wind and solar, other renewable energy technologies tend to produce at a constant rate over the day. Some, like biomass, are dispatchable. Others, like geothermal, generally run at

---

[10] It is possible that there are some minor additional maintenance costs associated with increasing generation, but these are very small, and renewables such as wind and solar are nearly always put ahead of dispatchable resources in the merit order of dispatch.

a constant rate all the time unless there is a maintenance shut-down. Of all major renewable technologies, only solar and wind are VERs.

Figure 2. Hourly Renewables Ouput Illustrative Day

Source: CAISO

These patterns of generation can be compared to the electricity *load*, which is the amount of electricity demanded at any period in time. Figure 3 plots the electricity load on March 31 in the CAISO electric grid for 2013 to 2016. The load is lowest during the night, ramps up in the morning, is relatively level throughout the day, and peaks in the evening. Wholesale electricity prices are based on both the supply and demand of electricity. In an electric grid with all dispatchable generation technologies, wholesale electricity prices would tend to follow the load over the day, with occasional spikes during steep ramps when higher marginal cost generation ("peaking plants") must be ramped up to meet those ramps. In contrast, retail prices are typically fixed and do not follow the load, as discussed above.

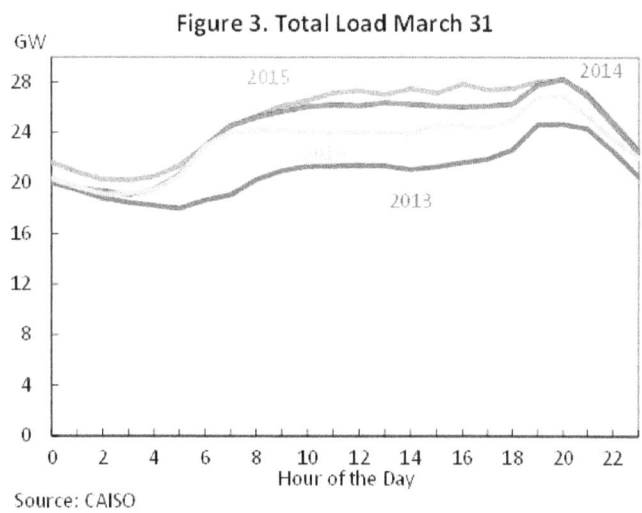

Source: CAISO

The nature of the supply side of an electricity market changes when there is a high penetration of wind and solar in the grid. Since the marginal cost (i.e., the additional cost of one more kWh of generation) of VERs such as wind and solar is (nearly) zero, they will generally be prioritized by the grid operator over dispatchable generation that has a positive marginal cost of generation.[11] It is generally less expensive to take the generation from a solar or wind facility when the facility has generation to offer than to take generation from a plant that has to pay for the fuel used to generate electricity. So, when the wind is blowing or sun is shining, the grid operator will usually take the electricity from wind or solar facilities, and thus there is a greater and cheaper supply of electricity at these times. The exception to this rule is during "curtailment" events, when generation from solar or wind resources exceeds energy demand, transmission limits prevent delivery of the electricity, or fossil generation units must be kept online to avoid costly start-ups and shut-downs. In these cases the resources are "curtailed," requiring that these facilities produce less energy than what they could have potentially provided.

The added supply to the grid is particularly beneficial when variable renewables are available at times of high load or high stress on the grid. For example, if the wind happens to blow during the evening hours, the cost of the displaced electricity from dispatchable resources is high. However, as shown in Figure 1, much of the generation for wind occurs at night outside of peak hours and much of the generation for solar occurs before the evening peak hours. At low levels of VER generation, these patterns of electricity generation are masked by variation in the load. At higher levels, these patterns can have a significant influence on the operation of the grid.

One example of how these higher levels can influence the grid is the case of West Texas wind. During occasional windy nights of the year when the load is low, there is so much supply of wind on the grid that it outstrips local demand. When there is insufficient transmission capacity, the areas with the greatest number of wind farms actually have negative wholesale electricity prices.[12] But West Texas is not alone. Another example is CAISO, where during the first quarter of 2016, there were numerous examples of negative prices in the late morning and early afternoon hours due to low load and high levels of renewable VER generation.[13]

A second example of how these higher levels can influence the grid is what has come to be known as the *duck curve*, so named because its shape resembles a duck. The duck curve is a plot of the load, just as in Figure 3, but after subtracting off VER generation. This is also often called the *net load*, since it is simply the load net of VER generation. The net load is the load that dispatchable generation will need to respond to. Recall that solar generation is the highest in the middle of the day, but load is often the highest in the evening. Thus, with enough solar generation, the net

---

[11] Many VERs even "self-schedule" their bids into the markets rather than having the grid operator schedule their bids, making them effectively "must-take" or "must-run" generation from a grid operator perspective.

[12] The production tax credit can sometimes help to make sales of wind generation at negative prices viable. In some cases, such sales may be viable because of inflexible contracts between the generators and the buyers of electricity. For instance, the buyer of the electricity (the load serving entity) may have a contract requiring them to make a per kWh payment regardless of the spot price of electricity, leading to disconnect between the firm deciding the output levels and the firm responsible for the negative prices.

[13] CAISO (2016c).

load could be lowest in the middle of the day and highest in the evening. The short time frame between the lowest and highest load of the day compresses the time needed to ramp up dispatchable electricity generation, leading to a very steep slope (i.e., "ramp rate") in the load curve between late afternoon and the evening.

Figure 4 presents an example of the duck curve in the CAISO electricity grid. It plots net generation for March 31 in 2012 and 2016, followed by projections for March 31 in 2017 and 2020 (the projections were made in 2013). In 2012, generation over the day followed a common pattern of the lowest load at night, higher load during the day, and higher load again in the late evening. Contrast this with the 2020 forecast, which represents a duck curve. The early morning hours have low net load (the duck's tail), the middle of the day has the lowest net load (the duck's belly), followed by an extremely steep ramp-up of generation (the duck's neck), to meet the highest net load of the day in the evening (the duck's head).

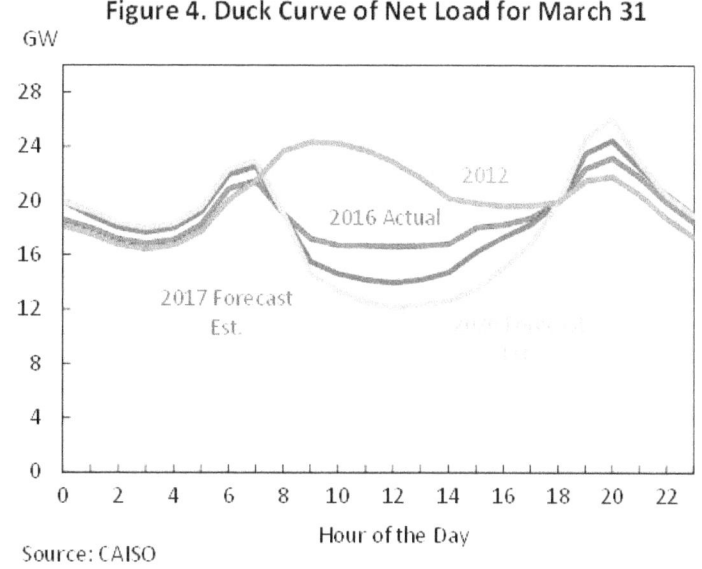

The actual net load on March 31, 2016 shows just the beginning of a duck curve. The lowest net load is in the middle of the day when the sun is shining, and there is a relatively steep ramp-up in the evening to meet the peak evening load. There is also a notable smaller peak in the morning while the sun is still low in the sky but activity has already begun for the day.

This discussion emphasizes that from a grid operator perspective, VER generation is (a) more uncertain than dispatchable generation and (b) may lead to steep ramps in the net load that traditionally must be met by dispatchable generation in the absence of energy storage. These two critical technical aspects of VER generation provide important opportunities for cost-effective and innovative smart markets and storage technologies to help integrate renewables onto the grid. To better understand the value streams possible in a future clean electricity grid that could support broader deployment of these solutions, we next discuss the costs of integrating VFR generation into the grid.

Technical Feasibility and Cost of Integration

There is a growing body of work examining the technical feasibility and cost of integrating higher levels of renewables onto the grid. Most of the focus of this literature has been on VERs, such as wind and solar, due to the unique characteristics of those resources, their rapid growth, and the substantial promise they hold as major contributors to a clean electricity system.[14]

As mentioned above, many energy analysts and academics had long claimed that there is a roughly 15 to 20 percent limit to the level of VER generation possible on an electricity grid. With a VER penetration higher than 20 percent, it was believed that managing the grid would become very difficult due to the uncertainty about when the generation would occur and challenges in ramping up and down dispatchable generation to offset the fluctuations in VER generation. However, as has been mentioned above, there are electric grids around the world today that are managing with extremely high levels of VER renewable penetration. These examples around the world have demonstrated that high levels of VER penetration are technically feasible with sufficient transmission capacity, connections with adjacent electricity grids, and careful management. However, technical feasibility alone does not speak to the cost of integration.

The cost of integrating renewable VERs onto the electric grid has been defined in many different ways. From an economics perspective, any definition must compare costs in a scenario with a given penetration of VERs to a baseline scenario with fewer (or no) VERs on the grid (generally the grid would be optimized in both scenarios). The most all-encompassing definition would include all of the differences in total system costs between these two scenarios over an extended period of time. This would include differences in the capital and fuel costs between the technologies, as well as changes in operating costs. A challenge with this broad approach is the difficulty in aggregating all of the potential costs and the possibility of double-counting.[15] Ideally, this approach could be extended to include any changes in consumer welfare from changes in electricity prices between the scenarios, as well as benefits from reduced emissions.

A more narrow definition focuses only on changes in the cost of operating the grid. This more narrow definition would include short-run costs such as (1) congestion costs from heavy use of certain transmission lines, (2) changes in the need to ramp up dispatchable generation, and (3) changes in the need for *ancillary services*, which are grid stability and reliability services. In the long run, this definition would also include the cost of any installed reserve generation capacity needed to back up the VER generation and perhaps additional congestion costs. This more narrow definition is useful because it focuses on services that have market prices and it reduces the possibility of double-counting. It can be examined separately from a comparison of the levelized cost of different generation technologies. The following addresses each of the categories of shorter-run integration costs.

---

[14] DOE (2015), IEA (2009), NREL (2011b), NREL (2015a).

[15] For example, including deferred fuel costs as negative values in the cost of integration, while at the same time using levelized cost of energy values (which incorporate fuel costs) is just one way in which double-counting could occur.

We can begin with congestion costs. If VER electricity generation is concentrated in a localized area, as may be the case with a major wind farm, then nearby transmission lines may be flooded with the VER generation and become congested. Grid operators then often impose "congestion charges" for use of the transmission lines that in equilibrium capture the marginal cost to the grid of using higher cost generation substitutes than ones that would be available if that line were not congested (e.g., ramping up a higher-cost generator accessible via less congested transmission lines). In the long-run these congestion costs can be addressed by adding new transmission capacity (or in some cases improving grid management). If such new transmission capacity is installed to facilitate VER generation, then these costs can be considered grid integration costs.[16]

As was seen above, a duck curve in net load would imply a very steep ramp up in generation between the late afternoon and the evening. Without smart markets and energy storage, grid operators can only accommodate this steep increase by retaining substantial quantities of quickly dispatchable generation resources (often highly emitting) or even turning off the generation from VERs. Many dispatchable energy resources take time to ramp up and it is costly to ramp them quickly. For example, most boilers in coal steam generation can form cracks from ramping up too quickly, increasing maintenance costs. This often leads to operators starting up generation well before it is actually needed in order to allow for a slower ramp-up, thus replacing lower-cost generation with higher-cost generation. While there are important exceptions, such as natural gas plants, in general the costs of dispatchable generation are higher during times of rapid ramping, leading to rapid increases in wholesale prices (i.e., the supply curve for dispatchable generation tends to be much steeper during these times). This is the fundamental concern of the duck curve. In fact, the higher wholesale prices during these times of steep ramping can be considered grid integration costs. One important caveat is that much recent research indicates that with sufficient transmission capacity and electric grids integrated over large regions, differences in wind speeds and solar irradiance across space can significantly dampen the steepness of the net load curve.[17] Indeed this is one important economic driver for the on-going movement towards integration of real-time and day-ahead western U.S. power markets.

Ancillary services are operations undertaken on the electricity grid essential for maintaining grid stability and reliability, such as voltage and frequency control (i.e., generators or other resources able to compensate if the voltage or frequency changes outside of the required range), regulation (i.e., dispatchable resources that can react in less than 5 minutes), spinning reserves (i.e., dispatchable resources ready to generate within 10 minutes), and other non-spinning operating reserves (i.e., dispatchable resources ready to start-up and generate within 30 minutes). Grid operators are likely to demand more ancillary services if VER generation increases uncertainty and variability in the net load. This would raise the price of ancillary services and the difference in the cost of ancillary services can be considered grid integration costs. For example, if additional capacity needs to be kept available due to increased uncertainty in broader fluctuations in the net load, this would raise the demand for spinning and non-spinning reserves. New technologies,

---

[16] Additional transmission capacity would have the benefit of smoothing the variability across a portfolio of VER facilities across space, due to the differences in wind speeds and solar irradiance across space.
[17] NREL (2011a), NREL (2012), Wolak (2015).

such as smart markets and energy storage, can be expected to help to dampen the increase in the price of ancillary services, by adding a new lower-cost substitute, as will be discussed in Section 4.

Figure 5 provides an example by showing the average net load and the average hourly price of an important ancillary service in CAISO in January 2016.[18] The highest ancillary service prices are during times of the greatest uncertainty in solar generation: early in the morning as the solar generation begins to ramp up and in the evening when the solar generation ramps down and electricity demand ramps up.

Figure 5. Avg. Market Clearing Price and Net Load

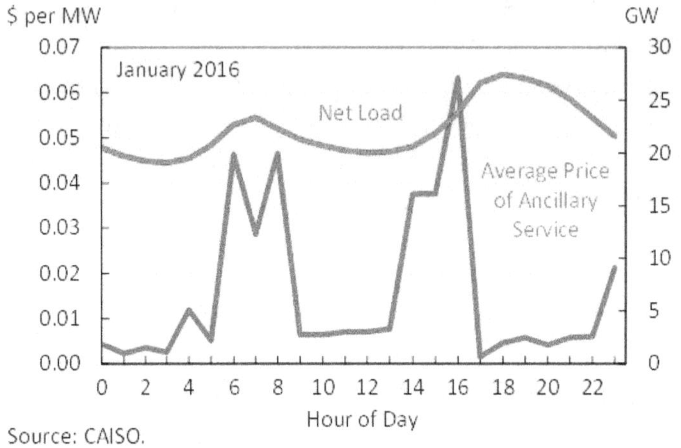

Source: CAISO.
Note: The ancillary service graphed is a service called "regulation mileage up".

Grid integration costs can be seen from two angles. On one hand, they are real economic costs from integrating VERs that may factor into decisions by policymakers or electric grid operators regarding electricity sources. On the other hand, they demonstrate the economic value of emerging technologies that can smooth the variability of the grid and reduce the steepness of the ramp in the duck curve. The next section provides evidence on the cost of integrating VERs into the electric grid and uncovers one of the more important value streams that could support deployment of emerging technologies.

---

[18] The ancillary service shown in the graph is the "regulation mileage up" service. Electric generators providing "regulation" service are paid to increase their generation at a very short notice in response to instructions from the grid operator. Though mileage payments represent a small portion of total payments received from providing regulation service, looking at mileage prices throughout hours of the day is instructive.

II. Evidence on the Cost of Integrating Variable Renewable Generation

As discussed above, one way that VERs such as wind and solar can increase the operating costs of the electric grid is by increasing demand for ancillary services used to reliably manage the grid, leading to higher ancillary service costs. Ancillary services are procured through markets, so we can examine market prices to understand the current costs and how they have changed over time as higher levels of VERs have been integrated into the grid. Although the exact ancillary service products that are traded through markets vary across grid operating areas, there can be markets for:

- Frequency response used for management of the frequency on the grid at 0 to 30 second intervals,[19]
- Regulation control used for providing a commitment to increase or decrease generation at 4 second to 5 minute intervals, and
- Spinning and non-spinning reserves used for commitments to increase generation at 10 – 105 minute intervals.

Voltage control and black start capability (i.e., ability to start from complete shut-down) are also often considered ancillary products, but are generally not procured through markets. The remainder of this section examines current and historical ancillary service markets to provide a benchmark for grid management costs and to discuss current and potential value streams could support broader deployment of smart markets and energy storage moving forward.

Current and Historical Ancillary Service Costs

We begin by exploring current and historical ancillary service costs at the market level. Since there has already been a dramatic increase in VER generation in many markets across the nation, one might expect ancillary service costs at the market level to have increased over time, and to be higher in markets that have a higher penetration of VERs.

Currently, ancillary service costs tend to be less than 3 percent of total costs per megawatt-hour (MWh) of load delivered, so they are a small, but not insignificant cost of electricity provision. But they vary greatly by market. Figure 6 plots average yearly ancillary service costs for managing the load in five different ancillary service markets across the nation over time against the level of variable energy penetration.[20] The average ancillary service costs per megawatt-hour (MWh) of load in an hour are calculated based on both the total spent on ancillary services (numerator) and the load procured (denominator), and thus may change over time if either the total ancillary service cost changes or the load changes.

---

[19] Currently there are no frequency response markets in the United States, but CAISO is considering establishing one.
[20] There are seven ISOs in the United States: PJM, CAISO, MISO, New England ISO (ISO-NE), ERCOT, NY-SIO, and SPP. Ancillary service costs per MWh load were not available in the market monitoring report for NY-ISO and SPP. For a map of the ISO regions, see the Appendix.

Either looking across markets or within a market over time, it does not appear that a rising VER share—within the range observed—leads to rising annual average ancillary service costs. Some markets have much higher VER penetration and yet have lower average ancillary service costs for managing their load than other markets with much lower VER penetration. In some markets, such as PJM (a major ISO centered in the mid-Atlantic region) and New England ISO, average ancillary service costs increase over time with VER penetration increases. In other markets, such as ERCOT and CAISO, there is a negative relationship between ancillary service costs and VER penetration. While different markets define their products slightly differently, we can get a quick sense of the relationship by pooling points across the regions. In doing so, we find that annual average ancillary services costs actually appear to decrease as VER penetration is increasing; the unweighted Pearson correlation coefficient over all data points is -0.42.

Figure 6. Ancillary Service Costs & Variable Energy Penetration
Annual Avg. AS Cost (2009 $ / MWh)

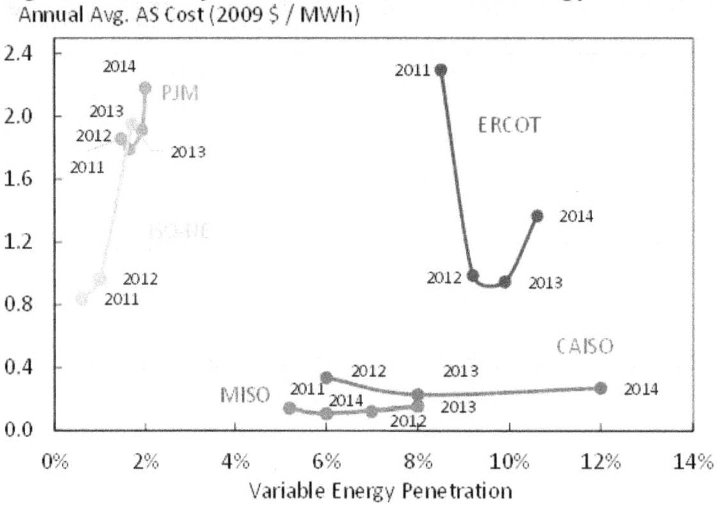

Source: Market Monitoring Reports for PJM, CAISO, MISO, ISO-NE, & ERCOT.

The lack of a positive relationship between current and historical average ancillary service costs and VER penetration has an important takeaway: the dramatic increase in renewable VER penetration up to the currently observed levels does not appear, at least from a first look, to have increased average ancillary service costs in a notable way. There may have been some increase in average ancillary service costs relative to a counterfactual world without VERs due to the increased demand for ancillary services, but such an increase appears to have been overwhelmed by other factors.

The factors driving ancillary service costs commonly cited in market monitoring reports are factors that influence ancillary service supply. Such supply factors can best be understood through the lens of opportunity costs by considering what the ancillary service supplier could have done otherwise if it did not supply ancillary services. When the opportunity cost of supplying ancillary services is high, then the supply of ancillary services decreases and the cost increases. For example, hydroelectric power often can provide ancillary services quite inexpensively. But during peak spring run-off when the water would be lost if it is not generating electricity, the opportunity cost of providing ancillary services–which may or may not generate electricity–is high, and thus ancillary service prices tend to increase in CAISO each spring since CAISO has a strong presence of hydropower. Similarly, if wholesale electricity prices are high, generators will prefer to generate electricity rather than offer ancillary services (see ERCOT in 2011 in Figure 6). On the flip side, markets with significant natural gas peaking capacity or hydroelectric capacity and low wholesale prices may have lower ancillary service costs because natural gas or hydropower can readily and inexpensively provide ancillary services. For example, the high share of hydroelectric and natural gas generation in CAISO may partly help explain why this region has lower ancillary costs than other regions despite a higher penetration of VERs.

Factors such as these have been dominating ancillary service costs so far, and the relatively low ancillary service costs in most markets have limited the incentive for innovation in new technologies. However, going forward, it is very possible that the demand for ancillary services from VERs will increase, as will their market value. In addition, we have already seen some evidence that ancillary service prices and costs increase as demand for these service increases.[21] We now turn to estimates in the academic literature for the cost of integrating higher levels of renewable VER generation into the electric grid.

Model Estimates of the Cost of Renewable Integration

There are several recent studies by the National Renewable Energy Laboratory (NREL), Lawrence Berkeley National Laboratory (LBNL), and others that use electric grid models to estimate the expected cost of integrating higher levels of renewable VER generation onto the grid.[22] The grid models used to estimate these costs tend to run at a short time scale and include generation, transmission, and distribution. They are specific to certain regions or electricity markets in the United States, for each market has a different set of attributes that can affect the cost of integration.

Figure 7 plots the estimates of average integration costs from different studies against the penetration of VERs. Note that the different studies use different definitions of integration costs, so one should only compare across them with caution. The recent historical prices of different ancillary service products are also included for another point of comparison (the comparison is again not perfect because the ancillary service prices are marginal costs, while the integration costs are average costs). The figure reveals that estimates of the cost of VER integration are remarkably similar, even to levels of VER penetration up to 30 percent. All of these estimates are also well within the range of recent prices of ancillary services and below many ancillary service prices. Note that the correct interpretation of integration cost estimates is that they are costs in addition to the ancillary service costs (and other costs) needed to maintain the grid without VERs. But it may not be surprising that they are within the range of historical ancillary service costs, since ancillary service costs capture the cost of dealing with current unexpected variation in the load that occurs often throughout the day.

---

[21] In February 20, 2016, CAISO increased regulation requirements. Following this increase, procurement costs for regulation capacity increased from less than $90,000 per day from January 1 – February 19, to almost $470,000 per day during February 20 through March 31 (CAISO 2016c).
[22] NREL (2011a), NREL (2015a), IEA (2009),

Figure 7. Ancillary Service Prices and Integration Costs per MWh Variable Generation

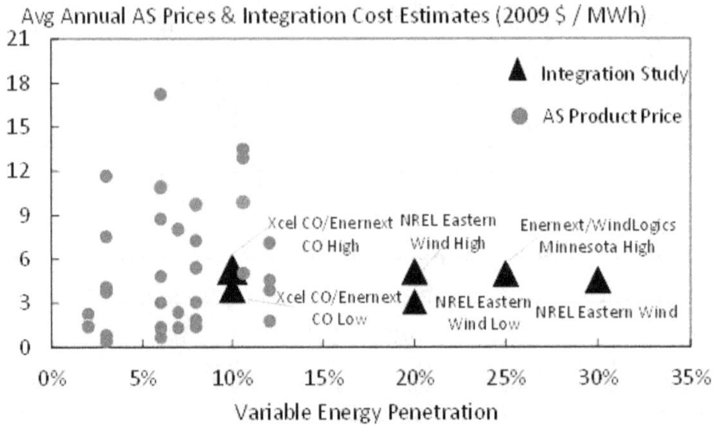

Avg Annual AS Prices & Integration Cost Estimates (2009 $ / MWh)

[Scatter plot showing AS Product Price (circles) and Integration Study (triangles) vs. Variable Energy Penetration (0%–35%). Labeled integration study points: Xcel CO/Enernext CO High (~10%), Xcel CO/Enernext CO Low (~10%), NREL Eastern Wind High (~20%), NREL Eastern Wind Low (~25%), Enernext/WindLogics Minnesota High (~25%), NREL Eastern Wind (~30%).]

Sources: Market Monitoring Reports from ISOs; NREL, Xcel Energy/Enernext, & Enernext/WindLogics.

Although the existing modeled estimates of average grid integration costs show a relatively flat supply curve with modest average costs of grid integration, the estimates only apply to specific markets and only up to 30 percent VER generation. The estimates are also averages that may mask changes in grid integration costs that are particularly important during certain times or in certain locations. At higher levels of VER generation, integration costs would be expected to increase meaningfully without private and public sector responses, such as new technological developments or innovations like smart markets and storage.

Evidence from Ancillary Service Markets

This section moves beyond aggregate statistics and uses hourly data from the CAISO market to examine how changes in the level of VER generation on the grid affect one type of ancillary service demand.[23] We focus on the ancillary service *regulation control*, which is used to provide a commitment to increase or decrease generation at 4 second to 5 minute intervals, as described above. Generators or other resources (e.g., energy storage) bidding into regulation markets can receive payments both for a day-ahead commitment to provide regulation services and for following instructions and actually providing those services if asked.[24]

In CAISO, generators or other resources bid their day-ahead commitments to provide regulation services into the regulation capacity market. The grid operator then chooses how much regulation capacity to procure (in MW) for a given hour in the next day based on the expected

---

[23] The data were downloaded for the CAISO_EXP region from the CAISO Oasis portal for the period May 2015 to May 2016 for day-ahead mileage requirements, day-ahead load forecast, and day-ahead output for solar and wind energy at the hourly level for the CAISO region.

[24] FERC Order 755 modified the payment structure for regulation services to embed a payment for the accuracy of following instructions.

grid conditions. This regulation capacity procurement does not typically change much over the day or even over seasons, so we generally observe only limited variation in the choice of capacity.

Once the capacity is procured, then at the time of usage the grid operator will send a set of instructions (usually at 4 second intervals) to the generator or resource, requesting increases or decreases of the power output. The sum of these instructions (in MW) over a given time period is called *regulation mileage*. For example, if in an hour a grid operator instructs the generator to increase output by 2 MW at second 8 and then later by 4 MW at second 40, the total mileage for that hour would be 6 MW.

In the CAISO market, the grid operator predicts the expected regulation mileage one day ahead to facilitate the procurement of capacity. We observe both the expected day-ahead regulation mileage at the hourly level and the actual regulation mileage instructions, although the actual instructions are seven day averages of hourly mileage. Examining how both the expected regulation mileage and the actual regulation mileage change with renewable VERs is useful for understanding how physical variability in the net load is related to renewable VER penetration.

Effect of variable generation on expected day-ahead regulation mileage

To begin, consider how much regulation mileage the grid operator will expect to procure in each hour of the next day. We would expect the operator to procure more ancillary services with greater load (e.g., due to concerns about an unexpected shut-down when generation capacity is near maximum capacity), greater VER generation, and steeper ramping of both load and VER generation. The actual day-ahead forecast of procurement is based on historical data, internal forecasts, and a variety of other factors, but we can model this procurement with the following reduced form:

$$AS_t = \alpha + \beta_1 Load_t + \beta_2 AbsSlopeLoad_t + \beta_3 VER_t + \beta_4 AbsSlopeVER_t + \mu_t^h + \mu_t^d + \mu_t^w + \epsilon_t, \quad (1)$$

where AS$_t$ is the forecasted quantity of regulation mileage in hour t (in MW), Load$_t$ represents load (in MW), VER$_t$ represents hourly VER generation (in MW), AbsSlopeLoad$_t$ represents the slope of load profile from hour t-1 to hour t, and AbsSlopeVER$_t$ represents the slope of output profile from hour t-1 to hour t. μ^h represents hour-of-the-day fixed effects, μ^d represents day-of-the-week fixed effects, and μ^w represents week-of-the-year fixed effects. ϵ_t is the residual, capturing all other unobserved factors impacting quantity of ancillary services procured.[25]

Table 1 presents the results of estimating several variants of the fixed effects model given in equation (1). Columns 1 and 2 present the results from model (1), while columns 3 and 4 further separate out VER generation by wind and solar technology to discern differences in how expected mileage changes with additional generation of each of the technologies. Columns 1 and 3 include

---

[25] To see why week-of-the-year fixed effects are included, consider a composite error term θ_t that represents all contingency events from dispatchable generators. These contingencies have seasonal patterns such as line downs during winter months and timed generator outages. Then, $\theta_t = \mu_t^w + \epsilon_t$ and we include week-of-the-year fixed effects to capture a portion of variation contained in the other contingencies category.

all of the fixed effects in equation (1), while columns 2 and 4 include only week-of-the-year and day-of-the-week fixed effects. Omitting hour-of-the-day fixed effects is useful for it includes variation across hours in the estimation, so it includes the effects of hourly ramping in the coefficients. With hour-of-the-day fixed effects, the coefficients are identified from variation in the data due to weather, seasons (the sun rising and setting at different times within a week), and other unexpected shocks that influence the load within a day.

Table 1. Results for Expected Day-Ahead Mileage

	(1) Total Mileage	(2) Total Mileage	(3) Total Mileage	(4) Total Mileage
VER Day-Ahead	0.016	0.0041		
	(0.0097)	(0.0050)		
VER Slope(AbsV)	0.11***	0.25***		
	(0.02)	(0.015)		
Load Day-Ahead	-0.0079*	0.045***	-0.010**	0.043***
	(0.0037)	(0.0023)	(0.0038)	(0.0023)
Load Slope(AbsV)	0.076***	0.10***	0.084***	0.10***
	(0.017)	(0.012)	(0.018)	(0.017)
Solar MW			0.044**	-0.0003
			(0.014)	(0.0050)
Wind MW			-0.0055	0.038**
			(0.013)	(0.013)
Solar Slope(AbsV)			0.11***	0.25***
			(0.026)	(0.014)
Wind Slope(AbsV)			-0.076	0.24*
			(0.097)	(0.093)
Hour-of-the-day indicators	Yes	No	Yes	No
Day-of-the-week indicators	Yes	Yes	Yes	Yes
Week-of-the-year indicators	Yes	Yes	Yes	Yes
Observations	8,780	8,780	8,780	8,780
R^2	0.36	0.31	0.36	0.32

Standard errors in parentheses
* $p < 0.05$, ** $p < 0.01$, *** $p < 0.001$

Columns 1 and 2 show broadly similar qualitative results. A key finding is that the slope of both the load and VER generation have a positive and highly statistically significant coefficient (statistically significantly different from zero at greater than a 99 percent confidence level), indicating that steeper ramping is positively associated with expected mileage. Grid operators may be able to partly anticipate ramping of load or VER generation, but view times with steeper slopes of net load or VER generation as more uncertain periods and thus will expect regulation mileage needs to be greater in order to cost-effectively ensure a reliable supply of electricity. The coefficient on the slope of the VER generation is greater in column 2 than column 1, which is to be expected since the hour-of-the-day fixed effects included in column 1 capture the consistent daily ramping of solar generation in the early morning and late evening. The coefficients in column 2 suggest that a 1 MW increase in the absolute value of the slope of load over an hour is associated with a roughly 0.1 MW increase in expected regulation mileage, while a 1 MW increase in the absolute value of the slope of VER generation is associated with a roughly 0.3 MW increase in expected regulation mileage. Thus, if we use our data to extrapolate from the coefficient estimate, a flattening in the slope of VER output from 9 to 10am (when the slope is the steepest) would reduce expected mileage in that hour by 14 percent. The coefficients on the level of VER generation and the load itself are not highly statistically significant (i.e., statistically significantly different from zero at greater than a 99 percent confidence level) in column 1, emphasizing the importance of ramping for the need for ancillary services.

Columns 3 and 4 parse out the mechanisms underlying the results in columns 1 and 2. The results in both of these columns are again generally qualitatively similar. The most notable result is that the coefficient on the slope of solar generation is positive and statistically significant, while the coefficient on the slope of wind is not statistically significant. This likely reflects the steep ramping of solar generation in the morning and evening (recall the duck curve), and the smoother ramping of wind, as shown in Figure 1. The coefficient on the solar slope in column 4 suggests that a 1 MW increase in the absolute value of the slope of load over an hour is associated with a 0.3 MW increase in expected regulation mileage, considering average solar slope in this hour, this corresponds to a 15 percent change in expected mileage for the hour between 9 and 10am. In column 3, the coefficient on the level of solar output is also statistically significant, which may correspond to the unpredictability of solar generation increasing unpredictability of net load. The effect is economically significant: a 1 MW increase in solar generation implies that the operator procures 0.04 MW more in expected regulation mileage, corresponding to a roughly 7 percent increase in expected regulation mileage from noon to 1pm when solar generation is at its peak.

It is also instructive to look at the fixed effects for the hour-of-the-day in the models in columns 1 and 3. The coefficient estimates on the dummy variables for the hour-of-the-day in both columns 1 and 3 show an intuitive pattern across the day in regulation mileage procurement (Figure 8). Expected mileage needs in the early- to mid-evening hours (hours ending in 16 and 21) are greater than earlier evening hours by 261 to 525 MW. This is all consistent with the daily ramping of solar generation described in Section 2, and in many ways matches the anticipated duck curve.

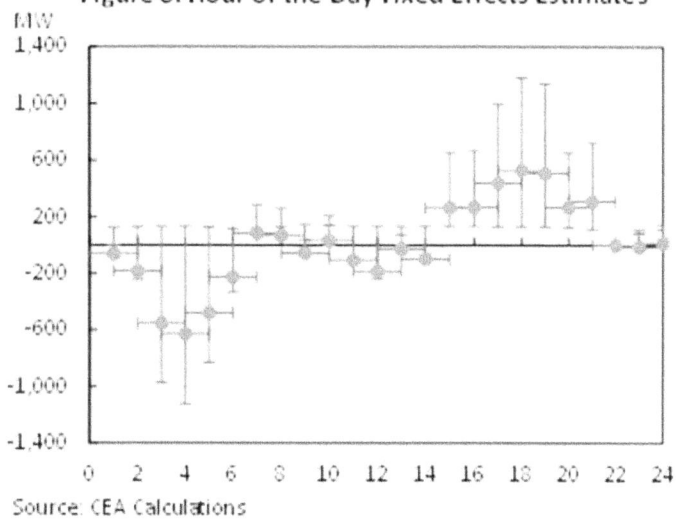

Figure 8. Hour of the Day Fixed Effects Estimates

Source: CEA Calculations

Effect of variable generation on actual regulation mileage

We can obtain further insight by examining actual regulation mileage from grid operations, rather than expected day-ahead regulation mileage. The mileage usage data contains 7-day averages of the actual usage in a given hour. In other words, for each hour t = 1...24, we observe average mileage use $\bar{M}_{td}$ for hour t for 365 days in our data set such that $\bar{M}_{td} = \frac{1}{7}\sum_{i=d-1}^{d-8} M_{ti}$. It follows that $7(\bar{M}_{td} - \bar{M}_{td-1}) = (M_{td-1} - M_{td-8})$ for t = 1...24. Let $\Delta \hat{d}$ represent the differenced observation of d-1 – d-8 within the same hour t. Let $\Delta M_{t\hat{d}} = (M_{td-1} - M_{td-8})$.

We can then model the change in actual mileage over the 7 days, with the following long differences model:

$$\Delta M_{t\hat{d}} = \pi_0 + \pi_1 \Delta Solar_{t\hat{d}} + \pi_2 \Delta Wind_{t\hat{d}} + \pi_3 \Delta Load_{t\hat{d}} + \phi_t^h + \phi_t^d + \phi_t^w + \varepsilon_t, \qquad (2)$$

where $\Delta Solar_{t\hat{d}}$ is the long difference in solar generation, $\Delta Wind_{t\hat{d}}$ is the long difference in wind generation, and $\Delta Load_{t\hat{d}}$ is the long difference in load. ϕ^h represents hour-of-the-day fixed effects, ϕ^d represents day-of-the-week fixed effects, and ϕ^w represents week-of-the-year fixed effects. ε_t is the residual, capturing all other unobserved factors impacting actual regulation mileage.

Table 2 shows the ordinary least squares results from estimating two variants of the long-differences model given by equation (2). Column 1 shows the results for the use of "regulation mileage up," which allows the grid operator to increase electricity generation at a short notice (i.e., the sum of a set of instructions to ramp up power generation). Column 2 shows the results for the actual use of "regulation mileage down," which allows the grid operator to decrease electricity generation at a short notice (i.e., ramp down power generation). The units for both regulation mileage up and regulation mileage down are MW observed in each hour.

Table 2. Results for Actual Regulation Mileage Up and Down

	(1) Regulation Down	(2) Regulation Up
Wind diff	-0.072	0.082*
	(0.039)	(0.038)
Solar diff	0.066*	-0.082*
	(0.027)	(0.035)
Load diff	-0.033**	0.030**
	(0.011)	(0.094)
Hour-of-the-day indicators	Yes	Yes
Day-of-the-week indicators	Yes	Yes
Week-of-the-year indicators	Yes	Yes
Observations	8,563	8,563
R^2	0.036	0.037

Robust standard errors clustered by week in parentheses.
* $p < 0.05$, ** $p < 0.01$, *** $p < 0.001$

The signs of the coefficients for the difference in load in Table 2 accord with intuition. A higher load compared to same hour last week requires more ramping up of resources (regulation mileage up) to meet demand, and a lower load relative to last week requires more the ramping down of resources (regulation mileage down). Both of these coefficients are highly statistically significant and are economically meaningful. A 1 MW increase in load over last week is associated with an increase in regulation mileage up by 0.03 MW, and a 1 MW decrease in load over last week is associated with an increase in regulation mileage down by 0.03 MW.

To put these findings in context, consider that over the first quarter of 2016 the largest hourly average deviation between forecasted and actual variable energy output occurred over 12 - 1pm, and equaled 582 MW.[26] Average deviation in load during this hour over the same time period was 541 MW. Taken together, an increase in load and solar output of these sizes respectively compared to the same hour in the prior week would lead to an increase in regulation mileage down usage of around 3 percent as compared to the average regulation mileage down usage in that hour.

The coefficients for solar generation match what might be expected for a generation resource that provides greater output than anticipated. Greater solar output relative to the previous week implies ramping down resources, while lower output relative to the previous week implies ramping up resources. These coefficients are not quite as statistically significant, but are economically significant. For example, a 1 MW increase in output over the previous week is associated with a 0.06 MW increase in regulation mileage down usage. Consider that over the first quarter of 2016 the largest hourly average deviation between forecasted and actual variable energy output occurred over 12 - 1pm, and equaled 582 MW. Then, an increase in solar output of this size would lead to an increase in regulation mileage down of around 2 percent of average regulation mileage down usage in that hour.

The coefficients for wind generation are more surprising: greater wind resources relative to the previous week in the same hour lead to lower usage of regulation down and higher usage of regulation up. One possible explanation for this is that while wind output does not exhibit steep ramping events like solar, minute to minute wind variability may be less easily forecastable. Then, higher wind output than expected exposes the grid to more unpredicted reductions in wind output, and thus requires greater use of ancillary services (e.g., greater use of "regulation up"). Lower wind levels than expected exposes the grid to more unpredicted increases in wind output, requiring the ramping down of resources (i.e., greater use of "regulation down").

Implications for Smart Markets and Storage

These results using data from CAISO may not apply uniformly across the country due to large differences in energy sources, generation patterns, and usage across grids. However, they shed light on the effects of wind and solar VER generation on physical grid management and potential value streams that could support deployment of smart markets and energy storage. While the

---

[26] Technically, this could be done at the sub-hourly level, but we use the hourly level for the purposes of illustration.

aggregate statistics show no discernable positive relationship between VER generation and ancillary service costs, our empirical results show that VER generation is already influencing the dispatch of regulation mileage. In particular, the empirical findings indicate that the steepness of ramping due to VER generation is as important, if not more important, than the level of VER generation for regulation mileage. The results are especially strong for solar generation, which relates closely to the steep ramp of the duck curve discussed above. The results suggest that as the duck curve becomes more pronounced, the impacts would become increasingly important.

This finding demonstrates both a need and opportunity for emerging technologies in storage and smart markets: technologies or approaches that reduce the steepness of the ramping of VER output can decrease ancillary procurement needs, lowering integration costs and facilitating integration into the grid. A quick thought exercise can highlight the value proposition. Suppose a technology or approach could reduce all of the steepness of VER output in the hour where the VER slope is the largest, from 9 to 10am, which has an average slope of 1657 MW over May 2015 to May 2016 for CAISO (the time frame of the dataset). Then, based on observed mileage and regulation needs over this time period, this technology or approach would provide a 14 percent savings in average mileage requirements in that hour. For illustrative purposes, suppose this 14 percent savings in mileage is accompanied by an equivalent savings in procurement of regulation capacity up and down. That is, assume that the amount of capacity that CAISO operators procured is proportional to the anticipated mileage, so that doubling mileage would double the capacity requirement. Even though mileage prices themselves are currently very low, regulation capacity costs can be substantial. April is an interesting month to examine in CAISO, since hydropower conditions create an upward pressure on ancillary services prices, with an average clearing price of regulation capacity in this hour of $17 per MW. Looking at average regulation capacity procurement and prices in April, we find that entirely flattening the steepness of the net load in this hour would save over $2,400 per day or almost $900,000 annually.

As discussed above, ancillary services are currently relatively inexpensive in CAISO due to high levels of hydropower and natural gas facilities. In other grid management regions, ancillary services prices are much higher. For example, in the PJM ISO in February 2015 the average market clearing price of regulation capacity between the hours of 5 and 6am is $123 per MW per hour.[27] Then, doing the same illustrative exercise with the CAISO data only taking this market clearing price, the same product for just that hour could earn a revenue stream of over $18,000 per day or roughly $6.6 million per year in regulation capacity payments alone. Going forward, with higher levels of renewable penetration, ancillary service costs would be expected to rise, increasing the potential value.

These findings underscore the opportunities that technologies and approaches to help with grid management present to facilitate deployment of higher penetration of renewable VERs onto the grid. The ability to smooth the steep ramping that can occur with VERs and reduce the uncertainty in generation will become increasingly valuable with higher levels of renewables. While assessing the exact cost impacts under future VER penetration scenarios is challenging

---

[27] For reference, the average 2015 regulation price in PJM for this hour ending 6am is $50 per MW per hour.

given the unknown future prices of regulation and variation in prices across markets, these results highlight that there are important value streams that have potential for growth moving forward. The next section will expand on these opportunities for emerging technologies and approaches.

III. Expanding Opportunities for Emerging Technologies

As the United States shifts towards increasing levels of VER penetration, the demand for ancillary services will be expected to rise, increasing the price of such services. This improves the value proposition for technologies and approaches to smooth fluctuations in generation, and in particular the steep ramps of the net load. Dispatchable generation can also provide additional ancillary services, but the cost of such resources will depend on the cost of the input fuels and the opportunity cost of providing these services, leaving open an opportunity for new technologies and approaches.

Recent market reforms by the Federal Energy Regulatory Commission (FERC) are also important for facilitating emerging technologies and approaches to support grid management. FERC Orders 755 in 2011 and 784 in 2013 reform the compensation structure for regulation services to improve the market valuation of resources providing grid support. Specifically, the orders require a payment structure that considers the speed and accuracy of service provision (e.g., with Order 755 frequency regulation ancillary service is compensated based on both capacity and performance). FERC Order 784 also introduces accounting and reporting rules that facilitate the ability of utilities to recover the costs of procuring energy storage capacity. Furthermore, FERC Order 792 from 2013 makes energy storage eligible for connection to the grid as an electricity generation source.[28] FERC also has an on-going request for information on barriers to the participation of storage in organized wholesale electricity markets, including energy, capacity, and ancillary service markets. Finally, FERC is taking comment on regulations regarding provision and compensation of frequency response in light of how stored energy resources can contribute by providing this service.

These market reforms, as well as additional reforms at the state or regional level, are already helping to facilitate the adoption of new technologies and approaches that will become all the more valuable with increasing levels of VER penetration. The remainder of this section discusses two major areas: smart markets and energy storage.

Smart Markets and the Smart End-user

As was discussed above, retail electricity prices are usually fixed over long periods of time, implying a highly inelastic demand for electricity. Advances in smart markets and, in particular, the smart end-user, hold promise to help flatten the net load curve, allowing for demand to respond to spikes in wholesale prices due to the intermittency of VERs or the higher costs of ramping to meet demand increases.

There is a very long history of electricity demand response to price signals going back to the turn of the 20th century, when engineers and utilities debated alternative pricing regimes that

---

[28] NREL (2016).

included charges at times of high demand and time-of-day differentiated rates.[29] In the 1970s, utilities began implementing load management programs to lower electricity system costs by reducing electricity demand during peak demand hours when wholesale prices are the highest.[30] There is a growing market for programs to manage electricity demand now, especially for commercial and industrial users. Yet, most retail electricity prices are still fixed and do not adjust with wholesale rates, eliminating any incentive for consumers to adjust patterns of consumption to the location and time-specific costs of providing electricity. At the same time, there has been considerable innovation in new enabling technologies, including communications technologies, smart meters, smart inverters, and other smart grid technologies.

These emerging technologies have the potential to greatly facilitate the response of electricity demand to changes in price (i.e., demand response). There are currently many types of demand response programs, which refer to any program that allows demand to change with either expected or actual wholesale price changes. Time-of-use electricity rates that follow a fixed schedule that is higher during times of peak wholesale prices and lower during the off-peak hours help to allow demand to change with expected wholesale prices. Other electricity rate schedules that allow some demand response include critical peak pricing rates, which charge higher prices for electricity during times of critical peak load on the system, and critical peak rebates, which provide rebates to customers who reduce their electricity usage during times of critical peak load. *Real-time pricing* (RTP) programs charge end-users the actual wholesale electricity price, providing an even more direct demand response to actual wholesale price changes.[31]

Another common demand response program is direct load control, where the utility or grid operator pays a customer for the right to turn down a customer appliance or system during a specified number of peak hours.[32] Further variants entail payments to customers for commitments to reduce electricity usage during times of critical peak load. FERC Order 745 has even made possible markets where customers bid in offers to reduce demand at a given price, allowing demand response to be treated as a supply resource.

Demand response programs improve economic efficiency by moving electricity prices closer to the true marginal cost of providing electricity at that location and moment in time. In doing so, these programs reduce demand when the quantity of electricity puts us on the steepest part of the supply curve for electricity. These gains can be substantial: Borenstein and Holland (2005) find welfare gains that are 5 to 10 percent of total wholesale energy costs from moving from flat pricing to real-time pricing. At the same time, demand response programs also improve fairness or equity: demand response can help avoid consumers who purchase electricity during off-peak times cross-subsidizing those who purchase during the peak. In the absence of real-time pricing at the retail level there are important economic efficiency and equity gains to be had from demand response.

---

[29] While perhaps ultimately unsuccessful, there was outspoken advocacy for time-of-day rates during these early electricity rate discussions (Hausman and Neufeld 1984).
[30] LBNL (2009).
[31] SmartGrid (2016a).
[32] SmartGrid (2016a).

Demand response programs can also help to capture key value streams that are expected to grow with further deployment of VERs. The first value stream comes about through simply reducing peak generation (i.e., "peak shaving"), which means that wholesale prices will not rise as high as they would without the demand response. This value is even more important with greater VER penetration due to occasional times when the sun is not shining and the wind is not blowing—times that would be expected to have high wholesale electricity prices. Demand response during these times can be quite valuable and is already being used by some utilities, although there is still room for further improvement in this area to nail down the correct monetization.

The second value stream also comes about through avoided transmission and distribution upgrades due to reduced peak generation. Transmission and distribution upgrades must be made based on forecasted increases in peak demand for electricity. Demand response can be used to avoid the cost of these upgrades. Utilities have also been using demand response for this purpose for decades, but are increasingly integrating it into their long-run planning, as evidenced by Distribution Resource Plans by utilities in California and the New York "Reforming Energy Vision" (REV) proceedings.[33] Continued expansion of smart markets and advanced communications is likely to make demand response all the more valuable going forward.

A third value stream involves reduced generation capacity. This value stream also relates to reducing peak generation, but also comes about through pricing uncertainty in supply. Utilities must ensure that they procure sufficient generation capacity to meet unexpected peak loads or unexpected shut-downs of some capacity. This requirement is called a *resource adequacy requirement* and is often met through markets where firms bid in spare capacity, called capacity markets. As the system transitions to higher levels of VERs, one would expect the amount of time dispatchable resources are run (i.e., their capacity factor) to decrease and older dispatchable capacity to be retired, both of which would contribute to a rise in the price of capacity.[34] Demand response programs are already trading on capacity markets in many regions, facilitated in part by FERC Order 745.

A fourth potential value stream comes about through smoothing the steepness of the ramp in net load, which can be expected to steepen with greater penetration of VERs (e.g., the duck curve due to solar generation). With a steeper net load, there are additional long-run system costs, such as from increased maintenance of dispatchable resources and transmission and distribution equipment. These are often not currently valued directly. However, as seen above in Figure 5, ancillary services costs tend to be higher during times of steeper changes in the load. As the duck curve becomes more pronounced with greater penetration of VERs, one would expect ancillary service prices to rise as the demand for ancillary services increases. Again, due in part to FERC orders, demand response and other flexible ramping products may be able to monetize this value stream.

---

[33] UtilityDive (2015).
[34] Mount et al. (2010).

A recent study by the Lawrence Berkeley National Laboratory (LBNL 2016) quantifies the potential amount of economic demand response available in California in future years. The focus of the study is on demand response bidding into the day-ahead electricity market and market for capacity at expected peak times when prices are high. The study finds that around 6 gigawatts (GW) of demand response is available in California by 2025 at an annual levelized cost of less than $200 per kW of demand response capacity available over a year. LBNL notes that $200 per kW is particularly relevant in this market because it is the cost of the relevant alternative for providing new capacity (e.g., delivered natural gas).[35] As this analysis does not consider value streams from the provision of flexible grid services, the quantity of demand response available at this price is likely to be an underestimate.[36]

Smart meters, as well as other devices to provide information to customers (e.g., "energy orbs" that light up during times of peak prices), have been shown in randomized controlled trials to further help focus consumer attention on reducing energy use at times when prices are high, increasing the elasticity of demand for electricity.[37] Jessoe and Rapson (2014) examine the impacts of devices providing high-frequency information about electricity usage and prices to residential customers, and find that households with access to high-frequency information are 13 percentage points more responsive to demand response pricing events than those without access to this information.[38]

Developments in smart appliances may eventually lead to a "smart home" that allows consumers to shape energy consumption patterns—and adjust them during times of higher prices.[39] For example, PNNL (2012) analyzed savings from a line of GE appliances that can automatically reduce demand (i.e., turn down or off) based on information received from the grid. The results of the study indicate annual wholesale energy market savings for the average household with smart appliances in the range of $40 to $60 in the New York ISO and $30 to $40 in PJM. One smart appliance with great potential is the electric hot water heater, which could be designed to match when the heating of water occurs with how stressed the grid is, while at the same time ensuring that there is enough hot water for the household.

Electric vehicles can also act as a smart appliance that allows for demand response. This is possible if there is communication between the vehicle and the grid, allowing the vehicle to automatically forego charging based on grid conditions.[40] Shao et al. (2012) find that demand response can allow for electric vehicle charging without any increase in peak generation on

---

[35] There is a question of whether it is most relevant to compare the cost of demand response to the cost of new fossil generation capacity or the cost of existing fossil generation capacity. In the longer-run, as plant retirements occur and there is load growth, the cost of new generation capacity is the most relevant. In the shorter-run, it is less clear. $200 per kW over a year may also be high; gas turbines are assumed by DOE to cost roughly $100 per kW per year.
[36] LBNL (2016).
[37] Faruqui and Sergici (2011).
[38] Jessoe and Rapson (2014).
[39] E3 (2016), SmartGrid (2016c).
[40] Silver Spring Networks (2011, 2013).

residential distribution circuits, thereby avoiding distribution circuit upgrades that would otherwise be required with EV deployment.

New technologies can also help more closely match distributed generation, such as residential and commercial solar photovoltaic systems, to times of high prices. For example, smart inverters, which can communicate with the utility and perhaps other nearby smart inverters, hold promise for staggering when solar systems come online and offline. This would ease voltage management issues (e.g., when there is low voltage, solar systems shut off, which can potentially exacerbate the issue further and even lead to brownouts or blackouts), lessen maintenance issues from ramping, and increase the value of the solar generation to the grid.[41] For example, Germany's smart inverter requirements are said to allow for 40 percent more PV capacity on the same line and are much less expensive than upgrading distribution infrastructure.[42]

New approaches can also enable novel technologies for smart markets and demand response. Interconnection rules do not universally allow demand response and distributed generation to participate in energy, capacity, and ancillary service markets. More broadly, reforms to the utility model toward models based on infrastructure as a service, rather than the classic model based on returns from capital investments (e.g., selling electricity), as well as reforms to the electricity rate structure, can help align economic incentives to encourage investment in smart markets and demand response. New York State's REV proceedings are notable for attempting to find a set of regulations that spur utility investment in the next generation of technologies by positioning the utility as a "distributed system platform provider" (i.e., a provider of software and infrastructure to support distributed generation by customers). Also notable are the California proceedings aimed at integrating distributed energy resources through a distributed resources plan.

Energy Storage

Analysts have long recognized inexpensive large-scale storage of electricity as a valuable resource when matched with VER generation. Today energy storage provides only a small contribution to the management of the grid. Out of 1066 GW of total electric generation capacity in the United States, there is only about 22 GW of electricity storage capacity installed.[43] But, with impressive energy storage technology improvements and an increase in VER generation, there is considerable movement towards expanded electricity storage.

Energy storage technologies come in several forms. Pumped hydropower storage is the most prevalent storage technology in the United States, representing over 90 percent of total storage capacity.[44] Pumped hydropower storage uses energy to pump water to an upper reservoir and releases this water to a lower reservoir to generate electricity when it is needed.[45] While an

---

[41] DOE (2016).
[42] IEEE (2015).
[43] EIA (2016b), NREL (2016).
[44] NREL (2016). One could consider hydropower in general to be energy storage, since it is storing potential energy. However, it is not typically categorized as an energy storage technology.
[45] MWh (2009).

effective energy storage approach, pumped hydropower storage is highly location-dependent, may have harmful impacts on aquatic ecosystems, can be constrained by water flow, and may not be as well-suited for the fast responses needed to provide some key short-run ancillary services.

Thermal storage offers another viable storage opportunity. Thermal storage sited at the end use, such as a commercial building, works by using electricity when prices are lower to cool some medium, often a coolant or water. The cooled medium (e.g., ice) can then be used to reduce electricity demand from air conditioning systems during peak hours when electricity prices are higher. Similarly, an electric hot water heater could be used to heat more water when electricity prices are lower (e.g., off-peak at night), storing the energy in the hot water so that less heating will be needed when electricity prices are higher.

Another form of thermal storage, which is growing in use rapidly, is the pairing of concentrating solar power plants with systems, often including molten salt, that store the heat and can be used with a turbine to produce electricity for some time after the sun stops shining. These paired solar and energy storage plants, such as Solana or Crescent Dunes in California, can help to smooth the intermittency of solar generation and positively change the duck curve by still producing electricity during the darker evening hours. However, such facilities are relatively expensive, require a great deal of land, and are best suited for sunny desert locations, making them less suitable for many areas of the United States.[46]

Flywheels are another possible energy storage technology. Flywheels are rotating mechanical devices that can store rotational energy that can be released by applying torque to a load, reducing the speed. While flywheels have been used in industrial and other niche applications for over a century, major technological advances are still necessary for flywheels to be economically viable for large-scale energy storage for the electricity grid.

Much of the recent interest in energy storage technology has focused on battery electric storage, which uses chemical reactions to store and release electricity. This has been spurred in part by dramatic cost reductions in battery technology in recent years. For example, lithium-ion battery packs have declined in cost by about 14 percent annually between 2007 and 2014, from above $1,000 per kWh to around $410 per kWh.[47] Looking ahead, with the further scaling up of battery production for electric vehicles and in-home usage, battery costs are expected to continue to decline. GTM (2016) projects that the full installed system prices (i.e., costs for the batteries plus labor and other materials) for utility storage projects for energy applications will decrease by 21 to 27 percent between the first quarter of 2016 and 2018. EIA (2013) projects lithium-ion battery costs will drop to as low as $135 per kWh by 2035, but some analysts and companies claim that these costs are already well under $200 per kWh today.[48]

---

[46] EIA (2012b).
[47] Nykvist and Nilsoon (2015).
[48] EIA (2013), Cobb (2015).

Battery energy storage holds promise both as a tool to support customer-level demand response and as a grid-scale resource. Batteries sited "behind-the-meter" can allow customers to time their electricity consumption to reduce costs under a real-time pricing or other dynamic pricing approach that brings wholesale prices closer to retail prices. For example, the Tesla Powerwall is a home battery that allows consumers to save money on electricity bills by arbitraging the difference between times of higher and lower prices. Such home batteries could even be coupled with distributed solar generation to help flatten the evening peak by using stored solar energy during those high value hours. Electric vehicles also have a battery that through smart markets and new technologies can provide additional value streams. For example, Kempton and Tomic (2004) find that 2.8 million battery vehicles could provide enough regulation capacity to support 700 GW of wind capacity (although such battery cycling for regulation may reduce battery life).

As a grid-scale resource, battery energy storage has the potential to capture the same value streams as demand response and smart markets. A major difference is that because battery storage capacity is expensive, but well-suited for providing electricity to the grid very quickly, the first value stream for energy storage capacity is likely to be in providing ancillary services, which are often critical during steep and unpredictable ramps of net load. But battery energy storage can also help meet resource adequacy requirements by bidding into capacity markets. It can provide value through arbitrage of high and low wholesale prices during the day, which one would expect to be all the more important if the duck curve becomes more pronounced. Finally, battery energy storage can help reduce the need for transmission and distribution upgrades based on forecasted increases in peak demand. All of these value streams are likely to increase over time with greater penetration of VERs on the grid.

Recent studies have examined how valuable energy storage can be in electric grids with a high penetration of VERs. DOE (2016) shows that integrating solar PV becomes increasingly expensive with higher solar PV penetration levels due to times when there is more generation than demand, leading to curtailment (i.e., stopping electricity generation when there is too much VER generation). But, adding demand response and energy storage can reduce the costs. For example, at solar PV penetrations approaching 24 percent, adding demand response and energy storage can reduce the overall levelized cost of electricity for solar PV by roughly 3 cents per kWh through reductions in required curtailments. NREL (2015d) finds substantial benefits in coupling solar photovoltaics with energy storage. In particular, the study examines couplings of customer-sited solar photovoltaic systems with lithium-ion battery systems in Los Angeles and Knoxville, where availability of behind-the-meter storage allows customers to take advantage of different electricity rates across hours of the day. The study finds that when coupled optimally, the net present value of an investment in a grid-connected battery storage facility can be as low as $300 per kWh.

Yet monetizing all of the value streams from storage has proven to be a challenge. Currently, energy storage can be financed either through utility investments to help maintain system reliability and reduce the need for transmission or distribution upgrades or through bidding into electricity markets, such as wholesale, capacity, and ancillary service markets. Even when energy storage is being monetized through electricity markets, it has often not been possible for energy storage developers to capture value streams in multiple markets, due to regulatory or logistical considerations.[49]

Recent regulatory changes are making it more possible to monetize value streams from energy storage and at least 10 states have introduced policies or regulatory actions since 2010. Perhaps even more importantly, several FERC orders have improved the opportunities for storage. The 2013 FERC Order 792 added energy storage as a power source eligible for grid interconnection as a small generator, and the 2011 and 2013 FERC Orders 755 and 784 changed the valuation of energy storage in ancillary service markets by compensating storage for its accuracy and speed of providing generation—making it much more viable to monetize the value stream from bidding into ancillary service markets. Energy storage costs are currently higher than the value offered for ancillary services in most markets, but as the penetration of VERs continues to build demand for ancillary services, the regulatory structure increasingly allows further value streams to be monetized, and energy storage costs continue to drop, the long-run prospects for energy storage will continue to brighten.

---

[49] Sioshansi et al. (2012).

IV. Conclusions

This report reviews the fundamental economic and technical aspects of integrating renewable VERs into the grid. The unique characteristics of renewable VERs can increase the demand for ancillary services to help manage the uncertainty and variability inherent in these generation technologies. The steep ramp of the net load in the duck curve is just one well-known example. At current levels of renewable VER penetration, the additional costs to the grid appear small or moderate, but as penetration rises, the demand for ancillary services can be expected to increase, leading to higher ancillary service costs to manage the grid. This implies growing opportunities for technologies and approaches that can help manage the grid by allowing for rapid and coordinated responses to changes in net load and wholesale prices.

Recent advances in smart markets and energy storage hold promise as solutions that can support further increases in VER penetration. This report identified several value streams that developers of smart markets and energy storage may be able to monetize to support broader deployment, including ancillary services, capacity markets, deferred transmission and distribution upgrades, and arbitraging wholesale price differences. Many of these value streams will become all the more important with higher penetration of renewable VERs, and new technologies and approaches may enable developers to monetize many of these value streams.

However, for the potential of smart markets and storage to be realized, there must be a level playing field allowing these technologies and approaches to participate in various electricity markets and be compensated appropriately for the value they provide, including their speed and accuracy. Efforts to remove barriers to participation in electricity markets, as well as efforts to ensure appropriate compensation for the value provided to the grid can greatly facilitate the growth in these new promising technologies. Similarly, efforts to improve the transparency of electricity prices and the connection between wholesale prices and retail prices can facilitate consumer demand response. There is already considerable progress in these areas, with notable regulatory reforms both at the state level and at the federal level by FERC. Yet there is still room for considerable innovation in this space. The reforms in this area have thus far been restricted to a limited number of states, and FERC is actively moving forward in considering further reforms. Further, there are many potential reforms that have yet to be implemented, which could include additional flexible ramping products or capacity products that are traded on markets that new technologies can bid into. With continued regulatory and technological development, the opportunities outlined in this report can support a cost-effective transition to a future resilient and low-carbon electricity grid.

References

Bloomberg New Energy Finance (BNEF). 2016. "Electric Vehicles to be 35 % of Global Car Sales by 2040."

Borenstein, Severin and Stephen Holland. (2005). "On the Efficiency of Competitive Electricity Markets with Time-invariant Retail Prices." RAND Journal of Economics. 36(3): 469-493.

California Public Utilities Commission (CPUC). 2016 "California Renewables Portfolio Standard (RPS)." http://www.cpuc.ca.gov/RPS_Homepage/.

California Energy Commission (CEC). 2016. "Renewables Portfolio Standard." http://www.energy.ca.gov/portfolio/.

California Independent System Operator (CAISO). 2016a. Open Access Same-time Information System (OASIS). http://oasis.caiso.com/mrioasis/logon.do.

_____2009. "AS Procurement – Regulation." *Technical Bulletin 2009-12-02*.

_____2014. 2013 Annual Report on Market Issues & Performance. Department of Market Monitoring.

_____2015. 2014 Annual Report on Market Issues & Performance. Department of Market Monitoring.

_____2016b. "Market Optimization." *Business Practice Manual for Market Operations Version 43*.

_____2016c. "Q1 2016 Report on Market Issues and Performance."

Cavallo, A.J., Hock, S.M. and Smith, D.R., 1993. *Wind energy: resources, systems, and regional strategies. Renewable Energy—Sources for Fuel and Electricity*. Island Press, Washington, DC.

Cobb, Jeff. (2015). "GM Says Li-Ion Battery Cells Down to $145/kWh and Still Falling." *HybridCars.com*.

Council of Economic Advisers (CEA). 2016. "A Retrospective Assessment of Clean Energy Investments in the Recovery Act."

DeCarolis, Joseph and David Keith (2005). "The Costs of Wind's Variability: Is There a Threshold?" *Electricity Journal* 18(1): 69-77.

Electricity Reliability Council of Texas (ERCOT). 2016. "Wind Integration Report: 02/18/2016."

Energinet.dk. 2016. "New record-breaking year for Danish wind power." http://energinet.dk/EN/El/Nyheder/Sider/Dansk-vindstroem-slaar-igen-rekord-42-procent.aspx.

Energy+Environmental Economics (E3). 2016. "Full Value Tariff Design and Retail Rate Choices." Prepared for: New York State Energy Research and Development Authority and New York State Department of Public Service.

Energy Information Administration (EIA). 2012a. *Annual Energy Outlook (AEO)*. U.S. Department of Energy.

_____2012b. "Electricity storage: Location, location, location… and cost." *Today in Energy*.

_____2013. *Annual Energy Outlook 2013*.

_____2016a. *Monthly Energy Review (MER) May*.

_____2016b. *Electric Power Monthly – March 2016*.

Ernest Orlando Lawrence Berkeley National Laboratory (LBNL). 2009. *Demand Response in U.S. Electricity Markets: Empirical Evidence*.

_____2016. *2015 California Demand Response Potential Study: Interim Report on Phase I Results.*

Farmer, E. D., V. G. Newman, and P. H. Ashmole. 1980. "Economic and operational implications of a complex of wind-driven generators on a power system." Physical Science, Measurement and Instrumentation, *Management and Education-Reviews, IEE Proceedings A* 127(5): 289-295.

Faruqui, Ahmad and Sanem Sergici. (2011). "Dynamic Pricing of Electricity in the Mid-Atlantic Region: Econometric Results from the Baltimore Gas and Electric Company Experiment." *Journal of Regulatory Economics* 40: 82-109.

Hausman, W. and J. Neufeld. 1984. "Time-of-day Pricing in the U.S. Electric Power Industry at the Turn of the Century." *RAND Journal of Economics* 15(1): 116-126.

Hawaii State Energy Office. 2016. Securing the Renewable Future. http://energy.hawaii.gov/renewable-energy. House of Representatives States of Hawaii (State of Hawaii). 2015. H.B. NO. 623.

IEA Wind Task 25 (IEA). 2009. "Design and Operation of Power Systems with Large Amounts of Wind Power: Final report, IEA Wind Task 25, Phase one 2006 - 2008."

IEEE Spectrum (IEEE). 2015. "How Rooftop Solar Can Stabilize the Grid." Accessed 6 June 2016. http://spectrum.ieee.org/green-tech/solar/how-rooftop-solar-can-stabilize-the-grid.

International Renewable Energy Agency (IRENA). 2015. "Battery Storage for Renewables: Market Status and Technology Outlook."

Irish Wind Energy Association. 2016. "Wind Statistics." http://www.iwea.com/windstatistics.

Jessoe, Katrina, and David Rapson. 2014. "Knowledge is (Less) Power: Experimental Evidence from Residential Energy Use." *American Economic Review* 104 (4): 1417-1438.

Kirby, Brendan. 2007. "Ancillary Services: Technical and Commercial Insights." *Prepared for Wartsila.*

Kempton, Willett, and Jasna Tomić. 2005. "Vehicle-to-grid power implementation: From stabilizing the grid to supporting large-scale renewable energy." *Journal of Power Sources* 144(1): 280-294.

Malik, Noreen. (2015). "Lower-Cost Wind and Solar Will Drive Energy Storage Technology" *Bloomberg Technology.*

Monitoring Analytics, LLC. (2016). "2015 State of the Market Report for PJM."

_____ 2015. "2014 State of the Market Report for PJM."

Mount, Tim, et al. 2010. "The hidden system costs of wind generation in a deregulated electricity market." *System Sciences (HICSS), 2010 43rd Hawaii International Conference*. IEEE.

MWH. 2009. "Technical Analysis of Pumped Storage and Integration with Wind Power in the Pacific Northwest Final Report." Prepared for: U.S. Army Corps of Engineers Northwest Division Hydroelectric Design Center.

National Renewable Energy Laboratory (NREL). 2011a. "Cost-Causation and Integration Cost Analysis for Variable Generation." U.S. Department of Energy.

_____2011b. "Eastern Wind Integration and Transmission Study."

_____2012. "Integration of Variable Generation and Cost-Causation."

_____2014. "A Survey of State-Level Cost and Benefit Estimates of Renewable Portfolio Standards."

_____2015a. "Review and Status of Wind Integration and Transmission in the United States: Key Issues and Lessons Learned."

_____2015b. "Over-generation from Solar Energy in California: A Field Guide to the Duck Chart."

_____2015c. "Solar Energy and Capacity Value."

_____2015d. "Economic Analysis Case Studies of Battery Energy Storage with SAM."

_____2016. "Energy Storage: Possibilities for Expanding Electric Grid Flexibility." *Analysis Insights*.

New York State Energy Research and Development Authority (NYSERDA). 2016. "Governor Cuomo Announces $150 million Available for Renewable Energy Projects." Accessed 10 June 2016. http://www.nyserda.ny.gov/About/Newsroom/2016-Announcements/2016-04-21-Governor-Cuomo-Announces-Millions-Available-for-Renewable-Energy-Projects.

North American Electric Reliability Corporation (NERC). 2012. "NERC IVGTF Task 2.2 Report: Reliability Considerations for BA Communications with Increased Variable Generation."

Nykvist, Bjorn and Mans Nilsson. (2015). "Rapidly Falling Costs of Battery Packs for Electric Vehicles." *Nature Climate Change* 5(4): 329-332.

Pacific Northwest National Laboratory (PNNL). 2012. "Modeling GE Appliances: Cost Benefit Study of Smart Appliances in the Wholesale Energy, Frequency Regulation, and Spinning Reserve Markets." U.S. Department of Energy.

Palensky, Peter and Dietmar Dietrich. (2011). "Demand Side Management: Demand Response, Intelligent Energy Systems, and Smart Loads. *IEEE Transactions on Industrial Informatics* 7(3): 381-388.

Portuguese Renewable Energy Association (APREN). 2016. "Portugal Just Went 4 Straight Days Running Completely on Renewable Energy." http://www.apren.pt/pt/media/clipping/imprensa/pagina-4/.

Potomac Economics, Ltd. 2014. *2013 Assessment of the ISO New England Electricity Markets.*

_____2015. *2014 Assessment of the ISO New England Electricity Markets.*

_____2013. *2012 State of the Market Report for the ERCOT Wholesale Electricity Markets.*

_____2014. *2013 State of the Market Report for the ERCOT Wholesale Electricity Markets.*

_____2012. *2011 State of the Market Report for the MISO Electricity Markets.*

_____2013. *2012 State of the Market Report for the MISO Electricity Markets.*

_____2014. *2013 State of the Market Report for the MISO Electricity Markets.*

_____2015. *2014 State of the Market Report for the MISO Electricity Markets.*

_____2012. *2011 State of the Market Report for the New York ISO Markets.*

_____2013. *2012 State of the Market Report for the New York ISO Markets.*

_____2014. *2013 State of the Market Report for the New York ISO Markets.*

_____2015. *2014 State of the Market Report for the New York ISO Markets.*

RED Electrica de Espana. 2016. 'Balance eléctrico anual nacional." www.ree.es.

Shao, Shengnan, Manisa Pipattanasomporn, and Saifur Rahman. 2012. "Grid integration of electric vehicles and demand response with customer choice." *Smart Grid, IEEE Transactions*. 3(1): 543-550.

Silver Spring Networks (Silver Spring). 2013. "White Paper: How the Smart Gird Enables Utilities to Integrate Electric Vehicles."

_____2010. "White Paper: The Dollars – and Sense – of EV Smart Charging."

Sioshansi, Ramteen, Paul Denholm, and Thomas Jenkin. 2012. "Market and Policy Barriers to Deployment of Energy Storage" *Econ Energy Environ Policy J* 1(2): 47.

SmartGrid. 2016a. "Time Based Rate Programs". U.S. Department of Energy. Accessed 2 June 2016. https://www.smartgrid.gov/recovery_act/time_based_rate_programs.html.

_____2016b. "Advanced Metering Infrastructure and Customer Systems". U.S. Department of Energy. Accessed 2 June 2016. https://www.smartgrid.gov/recovery_act/deployment_status/ami_and_customer_systems.html.

_____2016c. "The Smart Home". U.S. Department of Energy. Accessed 2 June 2016. https://www.smartgrid.gov/the_smart_grid/smart_home.html.

SolarCity. 2016. "A Pathway to the Distributed Grid." *White Paper*.

Stark, Camila, Jacquelyn Pless, Jeffrey Logan, Ella Zhou, and Douglas Arent. 2015. "Renewable Energy: Insights for the Coming Decade." *NREL Technical Report 6A50-63604*.

State of Colorado. (2016). "Renewable Energy Standard." https://www.colorado.gov/pacific/energyoffice/renewable-energy-standard.

U.S. Department of Energy (DOE). 2015. *Wind Vision*.

_____U.S. Department of Energy (DOE). 2016. "One the Path to SunShot: Emerging Issues and Challenges in Integration Solar with the Distribution System." NREL. Sandia National Laboratories.

UtilityDive. 2016. "ERCOT sets another wind record, with over 14 GW serving 45% of system load."http://www.utilitydive.com/news/ercot-sets-another-wind-record-with-over-14-gw-serving-45-of-system-load/414315/.

_____2015. "How California's Biggest Utilities Plan to Integrate Distributed Resources." http://www.utilitydive.com/news/how-californias-biggest-utilities-plan-to-integrate-distributed-resources/401805/.

Wolak, Frank. 2015. "Mean versus Standard Deviation Trade-offs in Wind and Solar Energy Investments: The Case of California."

Xcel Energy. 2006. *Wind Integration Study for Public Service Company of Colorado.* Work performed by Excel Energy and EnerNex Corporation. Denver, Colorado; Knoxville, TN.

Appendix

Figure A1. Map of ISO Operating Regions

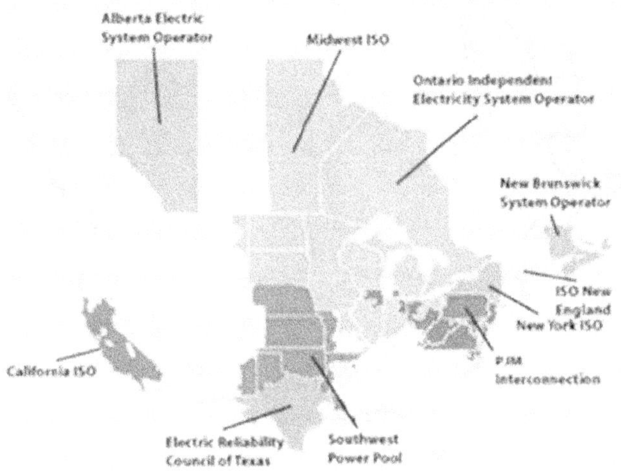

Source: ISO/RTO Council